Once upon a fall
in Sheridan, Wyoming...

there sprang a pumpkin patch.

Koltiska's Pride and Glory!

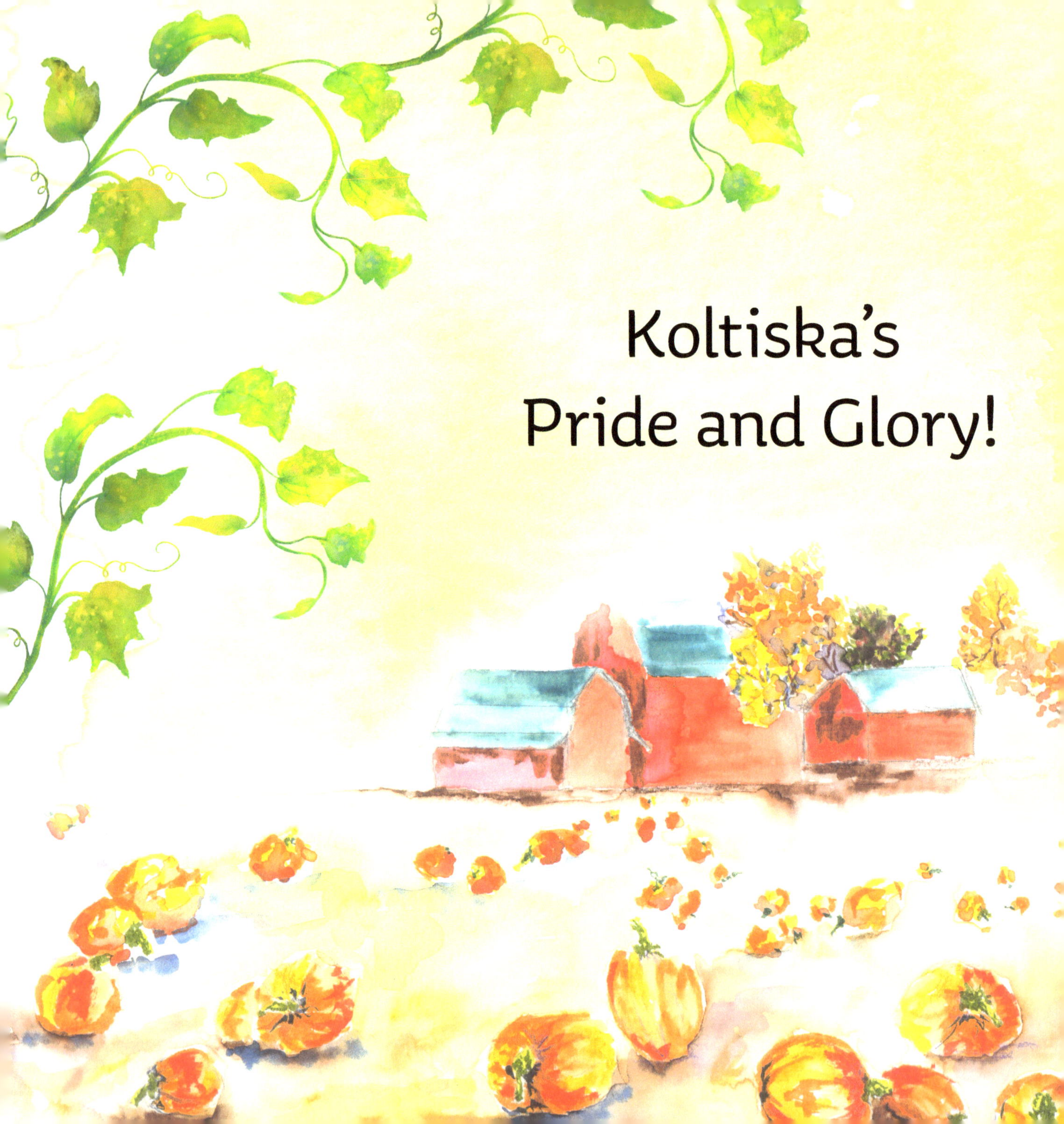

Gary & Vicki knew
how to have fun.

When kids came around,
there were candy and games,
and hayrides for everyone!

Hot chocolate was poured with marshmallows on top.

Or even hot cider for
Grandma and Pops!

It was out to the field
where the little kids ran...

to go find a pumpkin
as fast as they can!

Some were tall and skinny.
Some short and fat.

The pumpkins that is...
Not the "healthy" barn cat.

At the end of the day
when everyone went home,

they each had a pumpkin
to call their very own.

Come back and see them.
They're always here.

Let's find a pumpkin
in the fall of each year!

PUMPKIN FACTS!

Pumpkins are actually a fruit.
Many people think it should be our national fruit.

The earliest pumpkin pie made in America was quite different than the pumpkin pie we enjoy today. Pilgrims and early settlers made pumpkin pie by hollowing out a pumpkin, filling the shell with milk, honey, and spices and baking it.

Early settlers dried pumpkins' shells, cut it into strips and wove it into mats.

The world record weight for the largest pumpkin is 2,624 lbs!

Pumpkins were once considered a remedy for freckles and snakebites.

There are hundreds of varieties of pumpkins.

Pumpkins are good for your hearing!

Pumpkins are 80-90% water.

Each pumpkin contains about 500 seeds.

Every part of the pumpkin is edible, including the skin, leaves, flowers, and stem. Pumpkin and other squash blossoms can be eaten raw. They're also particularly tasty when lightly battered and fried!

The first Jack-o-Lanterns weren't made from pumpkins at all. They were made from turnips!

Over 1.5 billion pounds of pumpkins are grown every year in the United States.
(Equal to 30 of our Statue of Liberty)

Gary and Vicki Koltiska

Gary and Vicki Koltiska were neighbors of ours when we lived in Sheridan, Wyoming. Without them, Curtis would not have been able to start to build his dream of becoming a rancher.

In the mid 90's, Gary planted pumpkins and sold them to local stores. A few years later, they turned it into a pick your own pumpkin patch, slowly adding on fun features, one by one, like hayrides and hot chocolate. Now they are a legend in Sheridan, Wyoming, with buses full of kids coming to pick their own pumpkin each year.

Pumpkins are a brilliant crop - after the kids pick their pumpkins and the frost softens the remaining gourds, Gary turns his cows out on the ground who devour the mushy orange masses. His cattle get an amazing boost of natural health benefits, which they then deposit into the ground, and the cycle begins again in the spring when Gary plants the pumpkins again.

Gary is full of fun with a large laugh and is full of mischief. Vicki is a doting grandmother with a kind heart and gentle spirit. They have been an amazing inspiration to us as we have watched them toil together to make their dreams come true.

If you are ever near Sheridan, Wyoming in September or early October, make sure you make the extra side trip to Koltiska's Pumpkin Patch. It is worth the trip!

www.ingramcontent.com/pod-product-compliance
Lightning Source LLC
Chambersburg PA
CBHW050918210326
41597CB00003B/136